SCIENTIFIC AMERICAN EDUCATIONAL PUBLISHING

SCIENTIFIC AMERICAN INVESTIGATES GEOLOGY

WEATHERING AND EROSION

TARON LONGORIA

Published in 2025 by The Rosen Publishing Group
in association with Scientific American Educational Publishing
2544 Clinton Street, Buffalo NY 14224

Library of Congress Cataloging-in-Publication Data
Names: Longoria, Taron, author.
Title: Weathering and erosion / Taron Longoria.
Description: Buffalo, New York : Scientific American Educational Publishing, an imprint of Rosen Publishing, [2025] | Series: Scientific American investigates geology | Includes bibliographical references and index.
Identifiers: LCCN 2024009918 | ISBN 9781725351158 (library binding) | ISBN 9781725351141 (paperback) | ISBN 9781725351165 (ebook)
Subjects: LCSH: Erosion–Juvenile literature. | Weathering–Juvenile literature.
Classification: LCC QE571 .L68 2025 | DDC 551.3/02–dc23/eng/20240324
LC record available at https://lccn.loc.gov/2024009918

Portions of this work were originally authored by Maria Nelson and published as *Weathering and Erosion.*
All new material in this edition is authored by Taron Longoria.

Designer: Tanya Dellaccio Keeney
Editor: Brianna Propis

Photo credits: Cover GreenThumbShots/Shutterstock.com; p. 5 Lissj/Shutterstock.com; p. 7 AlicanA/Shutterstock.com; p. 9 Nature energy/Shutterstock.com; p. 11 trekandshoot/Shutterstock.com; p. 13 Thomas Pajot/Shutterstock.com; p. 15 saiko3p/Shutterstock.com; p. 17 Matthew J Thomas/Shutterstock.com; p. 19 Nicram Sabod/Shutterstock.com; p. 21 Jchannell/Shutterstock.com.

Printed in the United States of America

CPSIA compliance information: Batch #CSSA25. For Further Information contact Rosen Publishing at 1-800-237-9932.

CONTENTS

EVERYWHERE YOU LOOK

Have you ever noticed a large crack in the sidewalk in your neighborhood? Or maybe you've seen a gust of wind pick up and carry sand at the beach? If you have watched either of these happen, then you've seen weathering and erosion in action!

Weathering and erosion can have incredible outcomes. From the Grand Canyon in the United States to the Old Harry Rocks in England, these important **processes** can shape Earth and leave behind amazing landforms.

The Old Harry Rocks were created by thousands of years of weathering and erosion by the sea.

BREAKING IT DOWN

Although they usually work together, weathering and erosion are different processes. Weathering is the breakdown of rock into **sediment**.

Physical weathering is one way rock may break down. Sometimes called mechanical weathering, physical weathering changes a rock's size or shape. This can happen when water flows into open spaces in the rock, freezes, and causes the rock to crack. Other times, wind will carry and blow matter at a rock and, eventually, chip away little bits of it.

FUN FACT

ROCK MAY EXPAND IN HOT WEATHER AND SHRINK IN COLD WEATHER. EXTREME TEMPERATURE CHANGES CAN CAUSE ROCK TO WEAKEN AND BREAK MORE EASILY.

Weathering not only eats away at rock, but it can also cause statues and buildings to **deteriorate**.

CHANGING MAKEUP

Rocks are made up of **minerals**. Adding new minerals, **chemicals**, or gases to a rock may change its makeup and cause it to break down. This is known as chemical weathering. Rain is a common way chemical weathering happens. Rainwater mixes with chemicals and minerals as it falls. It then flows over and into rocks.

Biological weathering is the breakdown of rock because of the actions of living things. Animals may dig through soft rock. Some chemicals made by **bacteria** break down rock as well.

Mosses live on and feed on rocks constantly, which causes biological weathering. It changes the shape of the rocks!

FUN FACT

THE MINERALS THAT MAKE UP A ROCK, AS WELL AS THE ROCK'S OVERALL STRUCTURE, DETERMINE HOW IT WILL BE AFFECTED BY WEATHERING.

TRANSPORTING SEDIMENT

Once weathering creates sediment, the tiny bits of rock usually don't stay in one place. Sediment is often transported, or moved, to another place. When sediment is moved away from where it was created and **deposited** in a new place, it's called erosion.

There are many natural **agents** of erosion. Gravity, or the force that pulls everything toward Earth's center, is one major agent of erosion. It may work by itself or with other natural agents, such as wind or water, to cause erosion in different areas.

Landslides are an example of erosion caused by gravity. A landslide is the sudden movement of rocks and dirt down a hill or mountain.

LIQUID EROSION

Water isn't only important to weathering—it's part of erosion. It both breaks down rock into sediment and carries it away. This can happen in any body of water. For example, waves beat against lake and ocean shores, pulling bits of existing rock out into the water. Rivers wear away their banks over time too.

Rain causes erosion as well. Puddles form and begin to flow downhill, carrying soil and other matter. If a flood happens because of a lot of rainfall, more erosion will occur!

You may have seen erosion in action if you've ever been at the beach when the **tide** comes in. Shells and rocks are deposited on the sand, and that's erosion!

FUN FACT

LAKES, RIVERS, AND OCEANS HAVE LOTS OF SEDIMENT ON THEIR BOTTOM BECAUSE WATER PLAYS SUCH A BIG PART IN EROSION.

FLOAT AWAY

Another agent of erosion is wind. It blows around loose sand, soil, and other sediment. This often happens in very dry places, such as deserts. As wind moves some sediment, it creates even more sediment! Mountains and other landforms might be "sandblasted," or weathered by sand that's being transported by wind erosion.

Did you know glaciers cause erosion too? Glaciers are huge, slow-moving slabs of ice. They loosen and move rocks. Sediment can be transported as they move too.

As they move, glaciers melt and refreeze, which can lead to weathering and erosion too.

FUN FACT

SINCE SOIL EROSION CAN CAUSE HARM TO NATURE AND PROPERTY, PEOPLE TRY TO STOP IT FROM HAPPENING BY GROWING TREES AND PLANTS IN DRY AREAS TO HELP HOLD THE SOIL IN PLACE.

IMPORTANT STEPS

Why do weathering and erosion happen? These Earth processes are steps in the rock cycle, a scientific model that explains how rock breaks down and re-forms over time. A group of rocks called sedimentary rocks are created from the sediment that's made by weathering and transported by erosion.

As steps in the rock cycle, weathering and erosion are part of the creation of almost all the rocks on Earth, including those that form deep underground. They eventually reach Earth's surface over time.

Sometimes the words "weathering" and "erosion" are used to mean the same process. But remember, if sediment doesn't move, the process is weathering. If the sediment is moved and deposited, the process is erosion.

FUN FACT

WHEN SEDIMENTARY ROCK IS FORCED DEEPER UNDERGROUND, IT CAN BECOME THE TWO OTHER TYPES OF ROCK IN THE ROCK CYCLE: METAMORPHIC ROCK AND IGNEOUS ROCK.

DANGER AHEAD

Weathering and erosion are important parts of the rock cycle and the formation of rock on Earth. However, they can also be harmful. Erosion can hurt a farmer's crops. In dry years, wind blows away needed topsoil. Heavy rainfall may wash it away from plants that need it.

Rock loosened by weathering causes landslides and rockslides, fast-moving forms of erosion by gravity. These can fall on homes and block roads. When bad rainstorms occur, loose clay and mud can cause mudslides.

If you see a sign like this, that means you're in an area that has rockslides!

PLAYING A PART

Humans can also play a part in the increase of erosion. Clearing large areas of trees for building can cause soil erosion, which could make drinking water toxic. More erosion can also increase the amount of dust transported by wind—which can spread illnesses.

It is important to be aware of this and find ways to help slow erosion. Some people build rain gardens in their backyard to collect and clean extra rainwater. Others may plant special plants in their gardens that can survive lots of rainfall.

Rain gardens are a beautiful way you can help stop the dangerous effects of erosion.

HOW DOES WEATHERING AND EROSION HAPPEN?

CHEMICAL, PHYSICAL, AND BIOLOGICAL WEATHERING BREAK DOWN ROCK.

SEDIMENT IS CREATED.

EROSION BY WIND, WATER, GRAVITY, AND GLACIERS TRANSPORTS SEDIMENT.

SEDIMENT IS DEPOSITED.

GLOSSARY

agent: Something that produces an effect.

bacteria: Tiny creatures that can be seen only with a microscope.

chemical: Having to do with matter that can be mixed with other matter to cause changes. Also, the matter itself.

deposit: To let fall or sink.

deteriorate: To become worse over time.

mineral: Inorganic substances that make up Earth's rocks, sands, and soils.

physical: Having to do with matter.

process: The set of steps that move something forward.

sediment: Matter, such as stones and sand, that is carried onto land or into the water by wind, water, or land movement.

structure: The way an object looks, feels, and is arranged.

tide: The rising and falling of the surface of the ocean that happens twice a day.

FOR MORE INFORMATION

Books

Rogers, Marie. *Exploring Weathering and Erosion.* Buffalo, NY: PowerKids Press, 2022.

Sabelko, Rebecca. *Glaciers.* Minnetonka, MN: Bellwether Media, 2022.

Websites

Britannica Kids: Weathering
https://kids.britannica.com/kids/article/weathering/601271
Learn more about the process of weathering and look at some cool pictures!

GenerationGenius: What Is Erosion?
www.generationgenius.com/videolessons/changing-the-shape-of-land-video-for-kids/
Watch a video about how water and wind can change the shape of land, and test your knowledge with discussion questions!

INDEX